Things in Space

by Maryellen Gregoire

Consultant:
Adria F. Klein, Ph.D.
California State University, San Bernardino

capstone
classroom
Heinemann Raintree • Red Brick Learning
division of Capstone

A galaxy is in space.

Earth is in space.

The moon is in space.

The sun is in space.

Stars are in space.

Satellites are in space.

Even humans are in space!